black
misery

black misery

Langston Hughes

Illustrations by Arouni

Introduction by Jesse Jackson

Afterword by Robert G. O'Meally

Oxford University Press
New York • Oxford

Oxford University Press

Oxford New York Toronto Delhi Bombay Calcutta Madras
Karachi Kuala Lumpur Singapore Hong Kong Tokyo Nairobi
Dar es Salaam Cape Town Melbourne Auckland Madrid
and associated companies in
Berlin Ibadan

Library of Congress Cataloging-in-Publication Data

Hughes, Langston, 1902-1967.
Black Misery / Langston Hughes ; illustrations by Arouni ;
introduction by Jesse Jackson ; afterword by Robert G. O'Meally.
p.   cm. — (The Iona and Peter Opie library of children's literature)
1. Afro-American children—Juvenile humor.
[1. Afro-Americans—Wit and humor. 2. Prejudices—Wit and humor.]
I. Arouni, ill. II. Title. III. Series: Iona and Peter Opie library.
PS3515.U274B5   1994
818'5202—dc20   93-49590
CIP
ISBN 0-19-509114-0

2 3 4 5 6 7 8 9 0

Printed in the United States of America
on acid-free paper

INTRODUCTION
Jesse Jackson

Misery is sitting on the front of the bus because it's the first available seat. It's close to the driver. You want to see him switch gears and guide the bus. You can look out the front window. But the driver refuses to move, staring at you. The driver asks your mother to read the sign to you. "Colored seats in the rear. White seats in the front. Violators will be punished." And you have to walk past the other people who watch the spectacle and take your seat in the back of the bus. Your mother consoles you and explains to you things like, "The back of the bus will get to where the front is going. Besides, if there's a wreck, the people in the back are the least likely to get hurt."

Misery is being told, "Next year you're going to school." You look at your mind's image of school. One that's close to you. Where they have swings and the grass is green. And it looks like the school buildings in the books. Only when school does start, your mother takes you to school. You get to the school and get ready to turn up the sidewalk. She says, "You can't go to school here. It's for white children. You can't go here." You have to walk across to the other side of town. School is so crowded they have to have two shifts, 8 to 12 and 12:15 to 4. That school doesn't have any green grass. It isn't built like the other school.

These stories are true. They happened to me when I was a boy growing up in Greenville, South Carolina, in the early 1950s. They are the same sorts of experiences that Langston Hughes wrote of in this book, *Black Misery,* in the mid-1960s. Hughes had grown up in a period when the U.S. Supreme Court confirmed racist segregation laws in the southern states and Hollywood promoted stereotypes of African Americans as lazy and ignorant, as people who played the banjo and ate watermelon. Blacks were denied voting rights and endured the indignities of segregation on a daily basis. Hughes found that misery is a system of segregation that not only brands black children as "inferior." It also teaches white children hatred at an early age.

But when I grew up, I learned that joy is—celebration is—the something in our encounters, something outside us, that gives us the power to fight to change that system. That's when the term *freedom* explodes and something inside tells you that change is possible. And something inside tells you that you can be an instrument of that change. That's a joy and a peace that surpasses understanding—no matter how great the risk.

Langston Hughes achieved that joy—that celebration—through the power of words. He used the writer's imagination and command of language to fight the battle for justice. He skillfully laid bare the ironies and contradictions between the American creed—liberty and justice for all—and America's deeds.

When I was a boy, I read Langston Hughes. And then, when I was in college, I heard him give several lectures at Lincoln University in Pennsylvania. I was in such awe of him because I knew so much of him. And he was so impressive, a supremely confident man. He had that confidence because he could influence his environment. Because he was literate, because he was accomplished, because he was accepted. He had found a way to make the system give concessions to him. That's what literacy can do.

Have times changed since Langston Hughes wrote *Black Misery* in 1967? Yes. Young readers today may not remember the days when public places like schools and swimming pools were segregated by race. Or when black people could not vote. We no longer need the National Guard to let black children into public schools. Today, racial segregation is illegal. But there is still discrimination in America. Department store Santas are still, usually, white. The jobless rate for African Americans is double that of white Americans. The double standard in our educational system still tracks some youth to Yale while others are tracked to jail. The battle is not over. May the next generation find joy and celebration in overcoming the obstacles that remain.

Misery is when you heard
on the radio that the neighborhood
you live in is a slum but
you always thought it was home.

Misery is when the teacher
asked you who was
the Father of our Country and
you said, "Booker T. Washington."

Misery is when your pals
see Harry Belafonte walking
down the street and they holler,
"Look, there's Sidney Poitier."

Misery is when your white teacher
tells the class that all Negroes
can sing and you can't
even carry a tune.

Misery is when you learn
that you are not supposed
to like watermelon
but you do.

Misery is when the kid next door
has a party and invites
all the neighborhood but you.

Misery is when your
very best friend
calls you a name she really
didn't mean to call you at all.

Misery is when you call
your very best friend a name
you didn't mean to call her, either.

Misery is when nobody told you
the floorwalker would stop you
from riding up and down
the escalators 16 times
when mama is shopping.

Misery is when somebody meaning no harm called your little black dog "Nigger" and he just wagged his tail and wiggled.

Misery is when your mother said
considering the kind of family
your new friend Leroi has
even if he is white
you can't play with him.

Misery is when you have always
heard the old folks say Mississippi
is a place to be away from,
and on your first day
in a new school the kids ask,
"Are you from Mississippi?"

Misery is when you start
to play a game and someone
begins to count out
"Eenie, meenie, minie, mo...."

Misery is when you can see
all the other kids in the dark
but they claim they can't see you.

Misery is when you find out your bosom buddy can go in the swimming pool but you can't.

Misery is when the taxi cab won't stop for your mother and she says a bad word.

Misery is when you find out
Golden Glow Hair Curler
won't curl your hair at all.

Misery is when the colored actor
on the late, late show
bucks his eyes
at the wind shaking the shutters
as if he really believes in ghosts.

Misery is when you first realize
so many things bad
have black in them,
like black cats, black arts, blackball.

Misery is when your own mother won't let you play your new banjo in front of the <u>other</u> race.

Misery is when Uncle Joe gave you a button-up sweater and you wanted a slip-over that bunches at the bottom.

Misery is when you go to the Department Store before Christmas and find out that Santa is a white man.

Misery is when the only man
on the bus who is drunk
and talking out of his mind
is black.

Misery is when you wish
Daddy hadn't named
your dog Blackie.

Misery is when you start to help
an old white lady across the street
and she thinks
you're trying to snatch her purse.

Misery is when you come back
from the beach
proud of your suntan
and your pals don't even know
you've got one.

Misery is when you see that it
takes the whole National Guard
to get you into
the new integrated school.

Robert G. O'Meally

In April 1967 the publisher Paul Eriksson wrote to Langston Hughes proposing a book then only tentatively titled *Black Misery*. The idea was to follow through on the landslide success of Suzanne Heller's witty picture-and-caption books about (and ostensibly for) children called *Misery* (1964), *More Misery* (1965), and *Misery Loves Company* (1967). Like the present volume, each of Heller's books featured a simple line drawing and a single accompanying sentence to capture a moment from the period of childhood through adolescence in which public embarrassment or private chagrin reigned supreme. Every drawing showed a sad-mouthed child, every sentence offered a definition of his or her misery: "Misery is when you wear overalls and your straps fall in the toilet and you have to keep wearing them." "Misery is running away from home and not being allowed to cross the street." "Misery is a school dance which your parents have so generously offered to chaperone."

No wonder the Heller books were a smash. These stinging little vignettes of a middle-class child's life sold "by the bushel," Eriksson recalls—"over a quarter million copies before it was all over." The sales were not primarily to small children, as first expected, or to their parents. Perhaps kids as young as those depicted in the books were too close to the complexly painful scenes being depicted to react positively to them. Instead the *Misery* books sold best among college-age readers, primarily young women with enough distance from the books' characters to be able to look back and laugh at miseries they themselves had only just recently survived. "It has occurred to me," Eriksson wrote Hughes, "that there might well be great value and marketability in a similar book in the series which would be based entirely upon Negro 'miseries.'"

Hughes, who according to notes from April 1967 appeared at the top of Eriksson's list of possible authors for the projected little book, was the perfect choice. At 65 years of age, he was the best-known black writer in the United States, the one often introduced as "The Poet Laureate of Black America" or, to borrow one of his book titles, "Shakespeare in Harlem." Without question he was the most prolific African-American writer, and the one who had published literary works across the widest range of literary genres. He wrote poetry, short stories, novels, plays, essays, autobiographies, songs, and scripts for movies and television.

Working with his close friend Arna Bontemps, Hughes edited keystone anthologies and other teaching tools about life in black America, including *The Poetry of the Negro 1746–1949* (1949, 1970) and *The Book of American Negro Folklore* (1958). His writings first appeared in *The Brownies Book,* the children's magazine of the NAACP. One of his early collections of poems, *The Dream Keeper* (1932), was aimed at the children's market. He had written a clutch of other books specifically for children, including (with Bontemps) *Popo and Fifina* (1932) and several in an excellent series about first literary encounters: *The First Book of Negroes* (1952), *The First Book of Jazz* (1955), and *The First Book of Africa* (1960). Quite significantly for the present project, Hughes had contributed a superbly apt narrative text to a book entitled *The Sweet Flypaper of Life* (1955), featuring photographs by Roy Decarava.

Not only did Hughes know the territory of black America, but his work at its best turned on his genius for compressed evocative lyricism, the perfectly turned line. The poet Robert Hayden once spoke of the thrill and inspiration he felt when he read the wonderful poems in Hughes's first book, *The Weary Blues* (1926), with the words standing beautifully on the book's fine paper.

In some of his poetry, Hughes evoked the sound of blues music, and its terse and ironic attitudes toward the troubles of the world. In the title poem from *The Weary Blues,* he wrote:

> Droning a drowsy syncopated tune,
> Rocking back and forth to a mellow croon,
> I heard a Negro play.

Down on Lenox Avenue the other night
By the pale dull pallor of an old gas light
    He did a lazy sway....
    He did a lazy sway....
To the tune o' those Weary Blues.
With his ebony hands on each ivory key
He made that poor piano moan with melody.
    O Blues!
Swaying to and fro on his rickety stool
He played that sad raggy tune like a musical fool.
    Sweet Blues!
Coming from a black man's soul.
    O Blues!

Even when he did not employ the sounds and meters of the blues form as such, Hughes often wrote lines in which the point of view was bluely understated and comically wry. I quote the whole of "Little Lyric (Of Great Importance)": "I wish the rent / Was heaven sent."

Hughes's most popular works, the ones mentioned by the publisher in his letter about *Black Misery,* were those centering around the character whose name told something of his philosophy, Jesse B. Simple. Beginning in November 1942, Simple's history and deceptively simple, humorous reflections on life in Harlem (and, one might say, on what Ralph Ellison has called the "'harlemness' of the national human predicament") were reported in Hughes's nationally serialized newspaper column and then in a series of very popular books: *Simple Takes a Wife, Simple Stakes a Claim, Simple's Uncle Sam,* and *The Best of Simple.* Hughes explained, in his introduction to *The Best of*

Simple, that the idea for the character came to him when in a Harlem bar he asked a war plant worker what he made on his job:

> "Cranks," he answered.
>
> "What kind of cranks?"
>
> "Oh, man, I don't know what kind of cranks."
>
> "Well," I asked, "do they crank cars, trucks, buses, planes or what?"
>
> "I don't know what them cranks cranks," he said.
>
> At which his girl friend, a little annoyed, put in, "You've been working there long enough. By now you ought to know what them cranks crank."
>
> "Aw, woman," he said, "you know white folks don't tell colored folks what cranks cranks!"

Simple is Hughes's archetypal African American; he is life's underdog, mistreated and abused but nonetheless heroically undefeated. As the story "Feet Live Their Own Lives," also from *The Best of Simple,* shows, he is sharp-eyed, optimistic, unpretentious but eloquent, gutsy and indestructible:

> These feet of mine have stood in everything from soup lines to the draft board. They have supported everything from a packing trunk to a hungry woman. My feet have walked ten thousand miles running errands for white folks and another ten thousand trying to keep up with colored. My feet have stood before altars, at crap tables, bars, graves, kitchen doors, welfare windows, and social security railings.... In my time, I have been cut, stabbed, run over, hit by a car, tromped by a horse, robbed, fooled, deceived, double-crossed, dealt seconds, and mighty near blackmailed—but I am still here! I have been laid off, fired and not rehired, jim crowed, segregated, insulted, eliminated, locked in, locked out, left holding the

bag, and denied relief. I have been caught in the rain, caught in jails, caught short in my rent, and caught with the wrong woman—but I am still here!

As Robert Penn Warren has written, "With Jesse B. Simple, we may well have one of those creations, like Huck Finn by Mark Twain or George Babbit of Sinclair Lewis, who come from the common life into art and move from art back into life, to interpret and, somehow, ennoble life. And Simple resembles his creator, who, in his complexity and toughness, said of himself that he 'was a writer who wrote mostly because when I felt bad, writing kept me from feeling worse.'" Hughes's Simple demonstrated what his blues and bluesy poems demonstrated: a will to confront misery with heroic honesty and comic equipoise and to keep on stepping, to prevail.

At first Hughes hesitated to accept the offer to write *Black Misery* because the Black Revolution and the Black Arts movement of the 1960s had made his brand of humor seem old-fashioned. Although in his Simple column (as in his essays and other forms of writing) he had commented forthrightly, albeit not without laughter, on the varieties of continuing segregation, the civil rights marches, sit-ins, and monumental legislation, the racial bombings, the riots in Harlem and Watts, assassinations, and the myriad forms of black resistance—in 1965, after 23 years, Hughes announced that Simple was packing up to leave town; the column was over. He told a *New York Post* interviewer, "The racial climate has gotten so complicated and bitter that cheerful and ironic humor is less and less understandable to many people. A plain, gentle kind of humor can so easily turn people cantankerous, and you get so many ugly letters."

Rankled as he was by the humorless, violent rhetoric and the separatist ideology that characterized much youthful black expression of the mid-1960s, Hughes felt he understood it. He told a friend, "You must remember that sometimes we need the pendulum to swing way over in order to get things right. There are people who just have to be shaken up. The important thing, though, is to know when to stop." For his own part, Hughes had been exceedingly clear about what he felt to be the richest sources of inspiration for the African-American writer. In the words of Hughes biographer Arnold Rampersad:

> Hughes offered first an ethnic, then a universal principle, both of which he saw as intimately related. The first principle concerned Negritude, which was the same as American "soul." "*Soul* [wrote Hughes] is a synthesis of the essence of Negro folk art redistilled...particularly the old music and its flavor, the ancient basic beat out of Africa, the folk rhymes and Ashanti stories—expressed in contemporary ways so definitely and emotionally colored with the old, that it gives a distinctly 'Negro' flavor to today's music, painting or writing—or even to merely personal attitudes and daily conversation. *Soul* is contemporary Harlem's *negritude,* revealing to the Negro people and the world the beauty within themselves."
>
> He [Hughes] identified the universal principle: "If one may ascribe a prime function to any creative writing, it is I think, to affirm life, to yea-say the excitement of living in relation to the vast rhythms of the universe of which we are a part, to untie the riddles of the gutter in order to closer tie the knot between man and God. As to Negro writing and writers, one of our aims, it seems to me, should be to gather the strengths of our people in Africa and the Americas into a tapestry of words as strong as the bronzes of Benin, the memories of Songhay and Mele, the war cry of Chaka, the beat of the blues, and the *Uhuru* of African freedom, and give it to the world with pride and love, and...humanity and affection."

Ultimately, of course, Hughes did accept the *Black Misery* job. Perhaps he did so to help counterstate the sad and angry face so often associated with black protest and struggle; to show instead the African-American capacity (celebrated in the blues and in his own poetry) to survive hard times with good humor. To keep one's humanity intact despite trial and trouble.

Very much a work of the 1960s (but, alas, still quite relevant in our own time), most of its captions and pictures focus directly on the predicament of a black child adjusting to the new world of burgeoning integration and thus to new kinds of contacts with whites. *Black Misery* is fascinating as a document of that era's complexly hopeful but then again dismaying interracial relations. Most often the racism confronted here is of the invisible, super-subtle kind that lies embedded in the structure of the American language itself. Here what hurts is being defined by the then-current social science (and then pop journalistic) rhetoric of Negro pathology: "Misery is when you heard on the radio that the neighborhood you live in is a slum but you always thought it was home." The black kid feels it, too, when one of his white friends unwittingly uses a counting rhyme with a sharply racialized history: "Misery is when you start to play a game and someone begins to count out 'Eenie, meenie, minie, mo...'" (which according to one version, continues, "Catch a nigger by the toe").

To the black child of the 1960s, the world could seem a prison-house of color-coded language: "Misery is when you first realize so many things bad have black in them, like cats, black arts, blackball." Sometimes the race lingo comes out of the mouths of friends: "Misery is when your very best friend calls you a name she really didn't mean to call you at all." Sometimes it comes out of your own mouth: "Misery is when you call your very best friend a name you didn't mean to call her either." And at times the language of prejudice slices not along race lines but along the lines of class, with blacks not as victims but as perpetrators: "Misery is when your mother said considering the kind of family your new friend Leroi has even if he is white you can't play with him." Here we meet the radical Hughes who subtly reminds his readers that the problems of America are not just the ones of race against race but of class versus class.

The black children of *Black Misery* wince as they realize the stereotyped ways many whites see and don't see them and, in turn, the ways that they as blacks are expected to react. Perhaps the book's most successful caption defines misery as "when you learn that you are not supposed to like watermelon but you do." At times, good humor is tested to the limit: "Misery is when you start to help an old white lady across the street and she thinks you're trying to snatch her purse." *Black Misery*'s final caption makes explicit that the book's most profound subject is a black child's innocent effort to try to perform simple acts while a war is going on: "Misery is when you see that it takes the whole National Guard to get you into the new integrated school."

As if to underscore his universal intentions—articulated throughout his career—one of Hughes's *Black Misery* captions seems to have no racial overtones at all: "Misery is when Uncle Joe gave you a button-up sweater and you wanted a slip-over that bunches at the bottom." This is one of the book's several evidently nonracial squibs. But as Hughes knew, this caption, casually inserted among the more obviously racialized ones, assumes special black/white meanings, too, in context. Its twist lies in its ironically taking for granted that its nonblack audience realizes that many blacks operate in the middle-class world of Uncle Joe, the sweater as present, the youthful disappointment that the style isn't right. (If blacks are just as middle class as everybody else, what's all the fuss about anyhow?)

Taken by themselves, other pieces here also have "universal" significance—outside the strict parameters of race—but like the sweater caption, all have special force in the racialized zone of the whole sequence of *Black Misery*'s captions. "Misery is when nobody told you the floorwalker would stop you from riding up and down the escalators 16 times when mama is shopping." This is and was true for any child in any large store, but—as the illustration showing the tiny black child dwarfed by the enormous white man in the foreground emphasizes—the department store of the 1960s was a hotly contested, racialized zone. In the South it was a battleground where blacks had long been kept from eating at lunch counters and getting jobs and even from being waited on and trying on clothing. Child's play on the escalators, perhaps suggesting fast movement up beyond one's rightful place, could imply bigger trouble than the caption's quiet words make explicit.

Hughes died on May 22, 1967. He never got to finish this book, having written only 27 of the book's planned 45 captions. According to Eriksson, he was working on it in his hospital in his last days. At first, the publisher requested that the Hughes estate help to provide a few more entries, and Arna Bontemps offered to help by writing a few. But the additions did not seem to Eriksson to fit the book's tone and pattern. So the book went into production in 1969 with Hughes's pieces alone. Eriksson was right that the book seemed complete and whole without further additions. One snag was that in the end Suzanne Heller declined to do the illustrations, as she had done for her own *Misery* texts. She felt that as a white suburbanite she knew too little about the world Langston Hughes wrote about. So the press's designer, Lynette Logan (also white), known professionally as Arouni, volunteered and did a creditable job. The book came out in the summer of 1969, and it received solid notices and served as a reminder that the country had lost one of its poets of perfect pitch, the one Ellison had called (after the Hughes book title) "The Dream Keeper."

This book may work for certain children. But my guess is that its true audience will be those blacks like myself—as well as whites—who came of age in the 1960s and who were touched directly by the turbulent changes of those years, especially by the civil rights activism and legislation that cracked forever the walls of segregation. Here, to paraphrase Hughes, were some of the kinds of jokes blacks told among themselves. Like Hughes's characters, we endured impossible scenes similar to the ones chronicled here, and, like Hughes, we even managed an occasional smile at the craziness of it all, at what one Ellison character of this era calls the "United States of Jokeocracy." Maybe now, at last, looking back at these times, we can finally release the full extravagance of laughter which these painful, hopeful times of change deserve. Open the book to any page, be yanked back 20-something years, and give thanks to Arouni and especially to Brother Langston.

Langston Hughes was born in Joplin, Missouri, in 1902. He traveled all over the world—to Europe, Africa, Mexico, the Soviet Union—but his heart and home were in Harlem, where he was one of the most versatile writers of the artistic movement known as the Harlem Renaissance. Though known primarily as a poet, Hughes also wrote plays, essays, novels, short stories, and books for children. His writing is characterized by simplicity and realism and, as he once said, "people up today and down tomorrow, working this week and fired the next, beaten and baffled, but determined not to be wholly beaten." *Black Misery* was the last book that Langston Hughes wrote. He died in May 1967, while working on the manuscript.

Arouni is an artist who for many years lived in New York, where she illustrated and designed books. She now lives in Vence, France.

Jesse Jackson, president of the National Rainbow Coalition, has played a major role in virtually every movement for peace, civil rights, gender equality, and economic and social justice over the past three decades. As a young organizer in the Southern Christian Leadership Conference, he was assistant to the Reverend Martin Luther King, Jr. He went on to found Operation Breadbasket and Operation PUSH in Chicago, organizations involved in expanding economic and educational opportunities for minorities. Reverend Jackson is the recipient of the NAACP's Spingarn Medal and more than 40 honorary degrees.

Robert G. O'Meally is Zora Neale Hurston Professor of American Literature at Columbia University and previously taught English and Afro-American studies at Wesleyan University and Barnard College. He is the author of *The Craft of Ralph Ellison* and *Lady Day: Many Faces of the Lady* and editor of *Tales of the Congaree* by E. C. Adams and *New Essays on "Invisible Man."* Professor O'Meally is coeditor of *History and Memory in African American Culture* and *Critical Essays on Sterling A. Brown.*

THE IONA AND PETER OPIE LIBRARY OF CHILDREN'S LITERATURE

The Opie Library brings to a new generation an exceptional selection of children's literature, ranging from facsimiles and new editions of classic works to lost or forgotten treasures—some never before published—by eminent authors and illustrators. The series honors Iona and Peter Opie, the distinguished scholars and collectors of children's literature, continuing their lifelong mission to seek out and preserve the very best books for children.